# Quaternion Cosmos

Wardell Lindsay

© July 2006

## Contents

**1 Principles**     **2**
   1.1 Quaternion Numbers . . . . . . . . . . . . 8
   1.2 Quaternion Multiplication . . . . . . . . . 12
   1.3 Quaternion Geometry . . . . . . . . . . . 16
   1.4 Quaternion Group . . . . . . . . . . . . . 22
   1.5 Quaternion Differential Calculus . . . . . . 27

**2 Practice**     **31**
   2.1 Spacetime . . . . . . . . . . . . . . . . . . 34
   2.2 Ether . . . . . . . . . . . . . . . . . . . . 37
   2.3 Quantum Equations . . . . . . . . . . . . 40
   2.4 Electric Cosmos . . . . . . . . . . . . . . 45
   2.5 Relativity Equations . . . . . . . . . . . . 49

# 1 Principles

Quaternions are the numbers that control the Cosmos and most people may not have heard of Quaternions. Most math books don't mention them and most teachers don't teach them. This is why I wrote this book. Quaternions were discovered on October 16, 1843 by an Irish mathematician named William Rowan Hamilton. Hamilton was one of the first childhood prodigies in science. He was probably the greatest mathematician of his time. He had made the first spectacular prediction in physics 'conical refraction of light'. Hamilton predicted that light entering a crystal would come out not in a line, but would flare into a cone as it exited the crystal. No one had seen this, yet it happened just as Hamilton predicted. This prediction made Hamilton's reputation. He had many other contributions to mathematics including figuring out what what were the "imaginary numbers". Imaginary numbers were called "imaginary" because they were the square root of a negative number,

$$\sqrt{-1} \tag{1}$$

Square roots of negative numbers were strange but useful in solving equations. Hamilton, made "imaginary numbers" acceptable by showing them to be the result of a couple. Hamilton's couple is a number pair (a,b). Adding and subtracting couples, was ordinary, but multiplying couples was different:

$$(a,b)(c,d) = ((ac - bd), (ad + bc)) \qquad (2)$$

Now if 'a' and 'c' are zero, then the result would be (-bd, 0) a negative number from the rule for multiplication. Hamilton's couple was accepted and are called 'Complex' numbers and the imaginary symbol 'i', found a place in mathematics. Now we write the numbers as

$$(a + ib)(c + id) = (ac - bd) + i(ad + bc) \qquad (3)$$

Hamilton's pair or couple was the first development of the concept of vectors. The imaginary number had already made its appearance in trigonometry as an axis perpendicular to the real axis. Hamilton recognized this 'i' as a vector direction. His next step was to find three vectors to represent the 'x,y and z' of the real space. Then he would have a mathematical representation of real space. For 15 years Hamilton tried to use **'i, j and k' as three vectors** to rotate a line in the real space of 'x, y and z'. On October 16, 1843 his persistence paid off, as he was walking with his wife over a bridge, he had a Eureka! moment. He carved on the bridge the secret to mathematically rotating a line in real space:

$$\mathbf{i^2 = j^2 = k^2 = ijk = -1} \qquad (4)$$

Hamilton realized that four terms are needed; three vectors, 'i,j, k' and the scalar '1'. He called the four, Quaternions, from a verse in the Bible:

Acts 12:4 Peter to four quaternions of soldiers.

Quaternions are represented here by a scalar 'w' and three vectors 'ix + jy + kz'. The 'w' denotes a scalar and the 'r' denotes the three vectors.

$$Q = w + ix + jy + kz = Q_w + Q_r \qquad (5)$$

Hamilton in discovering quaternions, had discovered three vectors (i,j,k) all of which had negative squares and he clarified the need for the scalar term '1'. This was before Einstein's four dimensional "spacetime'. Nobody had a use for four dimensions, so only the three vectors found any acceptance. Vectors served a real need for directed forces in physics. Gravity and magnetism were directed forces, and a mathematics that could represent such directed forces was welcome. Many people adopted Hamilton's vectors, including James Maxwell. Maxwell made the greatest scientific prediction, even greater than Hamilton's, when he predicted that light was electricity! Maxwell was working in the emerging field of electricity and was looking for a way to represent directed fields, such as the magnetic field and the electric fields. Electric charges attracted and repelled each other as directed forces. The time was ripe for a mathematics that could do directions! Hamilton's quaternions were just such a mathematics.

In the 1800s, electricity was THE driving technology and vectors were driving electricity. Maxwell and others dropped the quaternion scalar using only the vector part in their work, typically an electric field, $E_r$:

$$E_r = iE_x + jE_y + kE_z \qquad (6)$$

Physicists had a problem with the minus sign when squaring Hamilton's vectors. Maxwell, complained , the negative sign gave him a minimum when he expected a maximum. For example when dropping a ball in the direction of gravity, the energy was negative not positive as Maxwell and other physicists expected.

Oliver Heaviside and J. Willard Gibbs, decided to get rid of this 'sign problem' and created vectors, I,J and K and changed the sign of the square to +1. This change was very popular and Gibb's Vector Analysis came to dominate the field and displace Hamilton's Quaternions. After the 1900, Hamilton's work became increasing difficult to find in textbooks and the mathematics literature. Today, the scientific community uses Gibb's Vectors with the square of the vector a positive one (+1).

I've left two pages after every section for your use as you read.

Thanks for buying this book.

Live and Learn,

Wardell Lindsay

*Truth is Beauty and Beauty is Truth*

*Truth is Beauty and Beauty is Truth*

## 1.1 Quaternion Numbers

I believe that understanding quaternions would reduce the learning load for all scientific education. Quaternions are only slightly more complicated than "Complex Numbers, which are a subset of quaternions, and quaternions can be applied to many more areas.

5x2=2x5=10. This is good old scalar arithmetic. Everybody was taught scalar numbers in school. This is an introduction to Quaternion Numbers.

A quaternion number consists of a scalar $Q_w = w$ and three vectors $Q_r = ix + jy + kz$:

$$Q = w + ix + jy + kz = Q_w + Q_r \qquad (7)$$

Every quaternion Q has a magnitude $|Q|$:

$$|Q| = \sqrt{w^2 + x^2 + y^2 + z^2} \qquad (8)$$

Quaternion Q represents an expansion or contraction by its magnitude and a rotation around its vector axis, 'q' of degree $\theta$.

$$Q = |Q|e^{q\theta} = |Q|(cos\theta + qsin\theta) \qquad (9)$$

where :

$$cos\theta = Q_w/|Q| \qquad (10)$$

Every quaternion has a negative:

$$-Q = -w - ix - jy - kz = -Q_w - Q_r \qquad (11)$$

Every quaternion has a conjugate $Q^*$:

$$Q^* = w - ix - jy - kz = Q_w - Q_r \qquad (12)$$

Quaternion addition and subtraction is commutative.

The sum of a quaternion and its conjugate is twice the scalar:

$$Q + Q^* = Q_w + Q_r + Q_w - Q_r = 2Q_w \qquad (13)$$

The Difference between a quaternion and its conjugate is twice the vector:

$$Q - Q^* = Q_w + Q_r - Q_w + Q_r = 2Q_r \qquad (14)$$

Every quaternion Q has a Norm $QQ^*$:

$$QQ^* = w^2 + x^2 + y^2 + z^2 = Q_w^2 + Q_r^2 = |Q|^2 \qquad (15)$$

Every quaternion Q has an inverse $Q^{-1}$:

$$Q^{-1} = Q^*/|Q|^2 = (w - ix - jy - kz)/|Q|^2 \qquad (16)$$

*Truth is Beauty and Beauty is Truth*

*Truth is Beauty and Beauty is Truth*

## 1.2 Quaternion Multiplication

Quaternion multiplication is different. Quaternions have four terms in them so multiplying four things by four things gives sixteen things. This is a lot more stuff. A better way to present it may be a Table.

$$Quaternion Multiplication Table$$

| $A_wB_w$ | $iA_wB_x$ | $jA_wB_y$ | $kA_wB_z$ |
|---|---|---|---|
| $iA_xB_w$ | $-1A_xB_x$ | $kA_xB_y$ | $-jA_xB_z$ |
| $jA_yB_w$ | $-kA_yB_x$ | $-1A_yB_y$ | $iA_yB_z$ |
| $kA_zB_w$ | $jA_zB_x$ | $-iA_zB_y$ | $-1A_zB_z$ |

(17)

$$AB = (A_wB_w - A_r \cdot B_r) + (A_wB_r + A_rB_w + A_r \times B_r) \quad (18)$$

$$AB = |A||B|(cosAcosB - sinAsinBcos(ab)) + \quad (19)$$
$$|A||B|(cosAsinBb + cosBsinAa + sinAsinBsin(ab)(a \times b)) \quad (20)$$

We can analyze the Table to correlate with the two equations following the Table which is the same quaternion product. The first is the Algebraic product and the second is the Trigonometric product of quaternions. The last term in the product is the non-commutative vector product term. Examining the table. we see the scalars in the left hand column and the top row. The upper left hand cell is $A_wB_w$, the product of the scalars. The left hand column contains the product of the vector and the scalar, $A_r B_w$. The top row contains the product of the scalar and the vector, $A_w B_r$. The diagonal of the table contains the quaternion scalar product, sometimes called the Trace of a matrix. The quaternion scalar product is the scalar product minus the vector scalar product. The diagonal is also the sum of cosines.

Vectors have two products the dot product, symbolized by the · and the cross product, symbolized by the ×. The vector dot product is a scalar, the vector cross product is a vector. 'Dot' and 'Cross' only refer to vector products. The cross product is the cause of non-commutativity. When the vectors are parallel the cross product is zero.

From the Trigonometry equation we see that the vector cross product is the last term containing the sine(ab). If the vectors 'a' and 'b' are parallel the sine is zero.

*Truth is Beauty and Beauty is Truth*

*Truth is Beauty and Beauty is Truth*

## 1.3 Quaternion Geometry

The Norm is the workhorse of Geometry, it underlies the Cosine Law ; the Matrix Determinant; the Characteristic Equation and the Pythagorean Theorem. The Cosine Law gives the square of the third side of a triangle. Let the third side be the difference between the two given sides, A and B.

$$C^2 = (A-B)(A-B)^* = (AA^* - (AB^* + BA^*) + BB^*) \tag{21}$$

In Quaternions the conjugate of the product is the reverse product of the conjugates:

$$BA^* = (AB^*)^* \tag{22}$$

This gives the following

$$C^2 = (AA^* - (AB^* + (AB^*)^*) + BB^*) = (AA^* - (2A \cdot B^*) + BB^*) \tag{23}$$

Which reduces to the Cosine Law

$$C^2 = (|A|^2 - 2|A||B|cos(ab) + |B|^2) \tag{24}$$

For the sum of two quaternions the minus sign becomes a positive sign. If cos(ab) is zero or the vector angle is a multiple of 90 degrees, we have the Pythagorean law.

The Characteristic Equation of Eigenvalue Fame is the NORM, replacing B with $\lambda$ and C being zero:

$$0 = (A - \lambda)(A - \lambda)^* = AA^* - \lambda(A + A^*) + (\lambda)^2 \quad (25)$$

$$0 = |A|^2 - 2A_w\lambda + (\lambda)^2 \quad (26)$$

Quaternions provide a mathematics for Geometry. The vector products are key here. The vector scalar product or dot product is a measure of perpendicularity. The dot product can result in a positive scalar, a zero and a negative scalar. Each of these have a different geometric meaning.

$$A_r \cdot B_r = -, 0, + \quad (27)$$

The square of a vector is negative if the two vectors are in the same direction. So in a Parity sense things in the same direction are negative and could be called negative Parity. Things pointing in the opposite directions have positive Parity. If the scalar is zero, the two vectors are perpendicular or have zero parity. This is a major mathematical aid to studying geometry and physics. Lines and forces can be analyzed mathematically for their physical orientation relative to each other. The sign of the scalar tells the orientation. The sign tells whether the forces are in the same direction(-) or opposite direction (+) or perpendicular (0). In Maxwell's experiment, the negative sign meant the force and the displacement were going in the same direction, which gives out energy, a condition of exergy. Lifting the ball against gravity would put energy in, and the scalar sign would be positive.

$$A_r \times B_r = -B_r \times A_r \qquad (28)$$

This vector product is called the 'cross' product. The result of this product is a vector, which is perpendicular to the two vectors in the cross product. The two vectors in the product form a plane and the result is a vector perpendicular to that plane. The area of the plane is proportional to the size of the two vectors multiplied by the sine of the angle between the vectors. The cross product is zero when the vectors are parallel! This is again a great aid to geometry. If the lines or forces are parallel, the cross product is zero.

This cross product is the reason why quaternions are not commutative!

$$ij = k \; ; \; jk = i \; ; \; ki = j \qquad (29)$$

Non- commutativity is the mathematical manifestation of the fact that rotations and spin (rotation around the axis) have directions, clockwise (cw) negative and counter-clockwise (ccw) positive.

$$ji = -k \; ; \; kj = -i \; ; \; ik = -j \qquad (30)$$

Rotate a vector counter clockwise , is called positive rotation and the result is pointing out of the clock or page. Rotate clockwise, is called negative rotation and the result is pointing into the clock.

The vectors i,j and k are responsible for non-commutativity and providing directional information mathematically. Most things in nature are non-commutative. If you turn a screw one way it goes in, if you turn it the opposite way it comes out. That's Non-commutativity!

*Truth is Beauty and Beauty is Truth*

*Truth is Beauty and Beauty is Truth*

## 1.4 Quaternion Group

Group Theory is something every educated person should understand. Group Theory shows how mathematics is at the heart of every system that we find useful to mankind.

Groups are important because, they underlie all mathematical and physical operations. Group Theory answers the question, can this equation be solved? Group Theory answers the question, is this process the same as this other process?

Group Theory got its start when people found it difficult to solve fifth degree equations. A mathematician named Galois, used Group Theory to show why, Fifth degree equations could not be solved by certain methods, but they could be solved by other methods. Modern engineering , including cell phone communications and Nuclear physics all are bound together by Group Theory. DNA is another area that will benefit from Group Theory. All kinds of useful communications are the result of Group Theory. Group Theory is the backbone of all systems, technical and non-technical.

Group Theory requires Four Features!

1. **Closure:** If i and j are members of the Group, ij=k is a member.
2. **Associativity:** i(ij) = (ii)j = -j
3. **Identity:** 1 such that, 1i = i1 = i
4. **Inverse:** $i^{-1}$ such that $i^{-1}i = -ii = 1$

Hamilton, was not trying to develop a Group, when he developed Quaternions. He was trying to rotate a line in three space with his vectors. It required Hamilton to define a Group to do this. Specifically, his attempts at rotation did not work until a valid Group had been defined. Quaternions had to be a group to do the job. The addition of the Identity member, a scalar (1), to the three vectors (i,j,k) formed the Group!

Remember, Gibbs changed the square of the his vectors to +1. Representing Gibb's Vectors by capitals I,J and K, we can see that Gibb's Vectors are not Associative.

$$I(IJ) = IK = -J \text{ but } (II)J = 1J = J \qquad (31)$$

Associativity holds for quaternion vectors:

$$i(ij) = ik = -j \text{ and } (ii)j = -1j = -j \qquad (32)$$

## The Cayley Table for the Quaternion Group

| 1  | -1 | i  | -i | j  | -j | k  | -k |
|----|----|----|----|----|----|----|----|
| -1 | 1  | -i | i  | -j | j  | -k | k  |
| i  | -i | -1 | 1  | k  | -k | -j | j  |
| -i | i  | 1  | -1 | -k | k  | j  | -j |
| j  | -j | -k | k  | -1 | 1  | i  | -i |
| -j | j  | k  | -k | 1  | -1 | -i | i  |
| k  | -k | j  | -j | -i | i  | -1 | 1  |
| -k | k  | -j | j  | i  | -i | 1  | -1 |

Unit Quaternions have a Norm of 1 and represent pure rotations.

$$U = |1|e^{u\theta} = |1|(cos\theta + usin\theta) \qquad (33)$$

There are of 8 Unit Singlets, the Quaternion Group:

$$\pm 1, \pm i, \pm j, \pm k \qquad (34)$$

There are 24 Unit Doublets:

$$\frac{\pm 1 \pm i}{\sqrt{2}}, \frac{\pm 1 \pm j}{\sqrt{2}}, \frac{\pm 1 \pm k}{\sqrt{2}}, \frac{\pm i \pm j}{\sqrt{2}}, \frac{\pm i \pm k}{\sqrt{2}}, \frac{\pm j \pm k}{\sqrt{2}} \qquad (35)$$

There are 32 Unit Triplets:

$$\frac{\pm 1 \pm i \pm j}{\sqrt{3}}, \frac{\pm 1 \pm i \pm k}{\sqrt{3}}, \frac{\pm 1 \pm j \pm k}{\sqrt{3}}, \frac{\pm i \pm j \pm k}{\sqrt{3}} \qquad (36)$$

There 16 Unit Quadruplets:

$$\frac{\pm 1 \pm i \pm j \pm k}{2} \qquad (37)$$

*Truth is Beauty and Beauty is Truth*

*Truth is Beauty and Beauty is Truth*

## 1.5 Quaternion Differential Calculus

<u>The First Quaternion Derivative: The Change Equation</u>

When Hamilton invented vectors, he also invented a vector Calculus. This vector Calculus involved a vector differential operator, he called Nabla after the shape of an Irish Harp $\nabla$. Hamilton did not invent a quaternion differential operator. When I was a studying electrical engineering, I wondered where Maxwell's Equations came from and my research led me to Hamilton's Quaternions and his Vector Calculus. I found it strange that Hamilton did not invent a quaternion Differential Operator, so I invented one by simply changing Leibniz's Scalar Time Derivative to a scalar space derivative using the speed of light $c$. I call this derivative Rad for radiation, symbolized by 'R':

$$R = \frac{\partial}{c\partial t} \qquad (38)$$

and adding this to to Hamilton's Nabla,

$$\nabla = \mathbf{i}\frac{\partial}{\partial x} + \mathbf{i}\frac{\partial}{\partial y} + \mathbf{k}\frac{\partial}{\partial z} \qquad (39)$$

created a Quaternion Differential Operator I call Xepera for Change, symbolized by $X$:

$$X = (\frac{\partial}{c\partial t} + \mathbf{i}\frac{\partial}{\partial x} + \mathbf{i}\frac{\partial}{\partial y} + \mathbf{k}\frac{\partial}{\partial z}) = R + \nabla \qquad (40)$$

### The Second Quaternion Derivative: Wave Equation

The second quaternion derivative is also a quaternion:

$$X^2 = (R^2 - \nabla^2) + 2R\nabla \qquad (41)$$

$$X^2 = (\frac{\partial^2}{c^2 \partial t^2} - \nabla^2) + 2\frac{\partial}{c\partial t}\nabla \qquad (42)$$

The Norm of the Differential is a scalar Differential:

$$XX^* = X^*X = R^2 + \nabla^2 = (\frac{\partial^2}{c^2 \partial t^2} + \nabla^2) \qquad (43)$$

The quaternion differentials are quaternions and multiply as quaternions. Quaternion Calculus is essential for Four dimensional Spacetime Quantum and Relativity Theory.

*Truth is Beauty and Beauty is Truth*

*Truth is Beauty and Beauty is Truth*

# 2 Practice

The Principles section of this book, is an exposition of quaternion mathematics. Mathematics can be applied to many different subjects and in different ways. This section is my view of physics. The focus here is not on all of physics but a few critical points that distinguish the major issues of the day.

Science is rightfully a conservative profession. However, in the case of quaternions, science has limited its own development by ignoring superior mathematics. Many in the scientific community still see mathematics as a calculating aid and miss the mathematical underpinning of physics. Much of our modern advance has come from the closeness with which physics and electrical engineering obey mathematical laws.

I touch on some of the critical issues in which quaternion mathematics provides insight into modern physics issues and let the mathematics speak. Hopefully, the first section of the book, has helped you learn the language of Quaternion Mathematics.

Many equations will look familiar, and you can know where they came from and what they mean. Some of th equations will be different and that too will be meaningful,when you know what is missing.

*Truth is Beauty and Beauty is Truth*

*Truth is Beauty and Beauty is Truth*

## 2.1 Spacetime

Spacetime is quaternionic, this means that our Cosmos has four dimensions, one scalar dimension (1) and three vector dimensions represented by (i,j,k).

A spacetime point can be represented by s:

$$s = ct+ix+jy+kz = ct+r = (ct)(1+r/ct) = (ct)(1+v/c) \tag{44}$$

where 'c' is the speed of light and 't' is time and 'v' is the velocity.

The Norm of s is:

$$ss^* = (ct)^2 + x^2 + y^2 + z^2 = (ct)^2(1+(v/c)^2) \tag{45}$$

The square of s is :

$$s^2 = (ct)^2((1-(v/c)^2) + 2(v/c)) \tag{46}$$

$$s^2 = ((ct)^2 - (x^2+y^2+z^2)) + 2(ct)(ix+jy+kz) \tag{47}$$

*Truth is Beauty and Beauty is Truth*

*Truth is Beauty and Beauty is Truth*

**2.2 Ether**

Maxwell believed that space contained an ether as a medium to propagate light. The existence of the ether has been disputed by many. I believe that the ether exists and is manifested by the so-called "free space resistance" to electromagnetic waves, 'z'. This resistance has a value around 375 Ohms. I believe that z is related to Planck's Constant and this relationship allows one to calculate the ether Quantum constituents, an Electric Charge and a Magnetic Charge.

$$z = W/C = 375\ Ohms \qquad (48)$$

$$h = WC = 2000/3 \times 10^{-36}\ joule - sec \qquad (49)$$

$$C = \sqrt{h/z} = 4/3\ atto\ coulombs \qquad (50)$$

$$W = \sqrt{hz} = 500\ atto\ webers \qquad (51)$$

There are disputes over the existence of the 'ether', but there are no disputes that Planck-Einstein's Radiant energy is 'hf' and the 'h' is the 'free space' constant that pervades the Cosmos. This constant exists and coexists with 'Maxwell's ether', the 'free space resistance', z.

*Truth is Beauty and Beauty is Truth*

*Truth is Beauty and Beauty is Truth*

## 2.3 Quantum Equations

Life is the essence of the Cosmos. There is a variable in the Cosmos that I call Life, that is a quaternion field and is the important variable for all of physics. Life is symbolized by 'L' and is related to Plancks constant 'h' and the speed of light 'c'. This variable has the units joule-meter. I chose this variable because the first derivative of this variable is 'work or energy' and the second derivative is force. Energy and force are key concepts in understanding the Cosmos.

The Life Quantum Constant has the value:

$$L_w = ch = 200 \times 10^{-27} \; joule\; meter(jm) = 1.25 \; uev. \tag{52}$$

$$L = L_w + iL_x + jL_y + kL_z = L_w + L_r = c(h_w + h_r) \tag{53}$$

Work is the first derivative of Life:

$$Work = XL = (R+\nabla)(L_w + L_r) = (\frac{\partial}{c\partial t} + \nabla)(L_w + L_r) \tag{54}$$

$$XL = (RL_w - \nabla \cdot \mathbf{L_r}) + (R\mathbf{L_r} + \nabla \times \mathbf{L_r} + \nabla L_w) \tag{55}$$

$$XL = (\frac{\partial L_w}{c\partial t} - \nabla \cdot \mathbf{L_r}) + (\frac{\partial \mathbf{L_r}}{c\partial t} + \nabla \times \mathbf{L_r} + \nabla L_w) \tag{56}$$

The Scalar Work Equation is Einstein's Photoelectric Equation and shows the "work function" to be the divergence of the life vector, $L_r$.

The boundary of Life is given by setting the first quaternion derivative to zero. This is the region where there is no longer any change. This is sometimes called the Conservation Equation, Continuity Equation, the Homeostasis Equation or the Homogeneous equation.

The scalar equation is the Quantum Equation of Planck's Constant, letting $L_w = ch$.

The vector equation is the Induction Equation familiar to Engineers in Electromagnetism. Newton's Law that "For every action there is an equal and opposite Re-action", is a statement of the Induction Equation.

The scalar Boundary Equation:

$$0 = (Rch - \nabla \cdot \mathbf{L_r}) \tag{57}$$

$$0 = (\frac{\partial h}{\partial t} - \nabla \cdot \mathbf{L_r}) \tag{58}$$

The vector Boundary Equation is The Quantum Induction Equation: Action Re-Action):

$$0 = (R\mathbf{L_r} + \nabla \times \mathbf{L_r} + \nabla L_w) \tag{59}$$

$$0 = (\frac{\partial \mathbf{h_r}}{\partial t} + \nabla \times \mathbf{L_r} + \nabla L_w) \tag{60}$$

Names for the differential terms:

$$RL_w = \frac{\partial h_w}{\partial t} \quad \textbf{Scalar Radiant} \qquad (61)$$

$$\nabla \cdot \mathbf{L_r} \quad \textbf{Divergence} \qquad (62)$$

$$RL_r = \frac{\partial \mathbf{h_r}}{\partial t} \quad \textbf{Vector Radiant} \qquad (63)$$

$$\nabla \times \mathbf{L_r} \quad \textbf{Curl} \qquad (64)$$

$$\nabla L_w \quad \textbf{Gradient} \qquad (65)$$

*Truth is Beauty and Beauty is Truth*

*Truth is Beauty and Beauty is Truth*

## 2.4 Electric Cosmos

The Quaternion Equations developed here, are the Conservation Equation(s) of the Electromagnetic Fields, where the first quaternion derivative is zero.

Setting the first quaternion derivative to zero is finding the Boundary or Conservation Condition, where there is no change. No change has many names, Boundary Condition; Homeostasis, Limit, Homogeneous, and Continuity Condition.

The fields are related to each other by two constants: $c$ the speed of light and $z$ the "free space impedance". These equations differ from "Maxwell's Equations" and I believe are truer. Experiments will decide.

$$E = cB = zH = zcD \qquad (66)$$

$$0 = XE = (R+\nabla)(E_w + E_r) = (\frac{\partial}{c\partial t} + \nabla)(E_w + E_r) \qquad (67)$$

$$0 = (RE_w - \nabla \cdot \mathbf{E_r}) + (R\mathbf{E_r} + \nabla \times \mathbf{E_r} + \nabla E_w) \qquad (68)$$

The traditional Equations involve two fields. This is accomplished by setting:

$$R = \frac{\partial}{c\partial t} \qquad (69)$$

$$0 = (\frac{\partial B_w}{\partial t} - \nabla \cdot \mathbf{E_r}) + (\frac{\partial \mathbf{B_r}}{\partial t} + \nabla \times \mathbf{E_r} + \nabla E_w) \quad (70)$$

The Electric Cosmos seems to be governed by this equation, the Conservation of the Electric Field. The scalar equations is the Continuity Equation and the quaternion expression of Ohms Current law. The Vector Equation is the Induction Equation and the quaternion equivalent of Ohms Voltage Law.

The electric field of the Sun warms the earth. The Sun power density is around 60 $Megawatts/m^2$. The Sun's Scalar Electric field is 150 kilovolt/m. The Sun's radiation wavelength is about 1 micron. The Sun's Radiant energy hf = RL = 1.25 electron volts.

The Induction Law is the Law of Motors and Generators. The Electric Cosmos consists of bodies operating as motors and generators. Currents flow throughout the Cosmos created by dynamos and consumed by bodies acting as motors. The Cosmos is alive with electricity. The ether is a reservoir of scalar quantum electric and magnetic charge.

*Truth is Beauty and Beauty is Truth*

*Truth is Beauty and Beauty is Truth*

## 2.5 Relativity Equations

Relativity focuses on 'gravitational force'and the curvature of spacetime. The second derivative of Life is a Force and a wave equation.

The Quaternion Wave Equation contains two waves. A scalar longitudinal wave (gravity) and a vector transverse wave (electromagnetism).

$$X^2 L = ((R^2 - \nabla^2) + 2R\nabla)L \qquad (71)$$

$$X^2 L = (R^2-\nabla^2)L_w - 2R\nabla \cdot L_r) + ((R^2-\nabla^2)L_r + 2R(\nabla \times L_r + \nabla L_w)) \qquad (72)$$

The scalar wave consists of the transport term called the D'Alembertian and has a divergence term indicating expansion. When the divergence is positive, the vector field("matter") expands. This could explain the so-called expansion of the universe by "dark matter". It is not "dark matter", it is the expansion of matter in response to Scalar Radiant Energy. Thermal energy is a common form of scalar radiant energy. This is a common effect. Heated solids expand and become liquids; heated liquids expand to become gases; heated gases expand to become plasma. The Cosmos will contract and release this Scalar Radiant energy. This effect is seen in Quantum events. The Quaternion Laws hold at all levels from the atom to the Cosmos. The Cosmos is Quaternionic.

The vector wave equation is transverse and is unidirectional. The vector wave has a curl indicative of transverse vibrations. In addition, the vector part contains the gradient of the scalar.

Gravity is a scalar force and the scalar wave equation is the "gravity" wave equation. From the "gravitational Constant", Gravity is a kind of Cosmos "pressure" of 1.44 Tera Newtons per meter squared. The gravitational force is transmitted through the ether as a scalar electric field of 402 Giga Volts/meter. The gravitational wavelength is 610 pico meters.

## *Truth is Beauty and Beauty is Truth*

Thank You for reading this far. Physics hasn't moved much since Einstein's Time. Much of this is due to a failure to appreciate the four dimensionality of the Cosmos and having a mathematics to deal with it. Understanding Quaternion Mathematics will give you insight into the way the Cosmos works.

Quaternions unites and simplifies The Physics of the Cosmos.

Truth is Beauty and Beauty is Truth.

Wardell Lindsay

*Truth is Beauty and Beauty is Truth*

*Truth is Beauty and Beauty is Truth*

*Truth is Beauty and Beauty is Truth*

*Truth is Beauty and Beauty is Truth*

Printed in Great Britain
by Amazon